Cambridge Primary
Mathematics

Second Edition
Workbook 2

Catherine Casey
Steph King
Josh Lury

Series editors:
Paul Broadbent
Mike Askew

Boost

HODDER
EDUCATION
AN HACHETTE UK COMPANY

Cambridge International copyright material in this publication is reproduced under licence and remains the intellectual property of Cambridge Assessment International Education.

Third-party websites and resources referred to in this publication have not been endorsed by Cambridge Assessment International Education.

Registered Cambridge International Schools benefit from high-quality programmes, assessments and a wide range of support so that teachers can effectively deliver Cambridge Primary. Visit www.cambridgeinternational.org/primary to find out more.

Acknowledgements

The Publishers would like to thank the following for permission to reproduce copyright material.

Photo credits
p. 8 *tl, cr,* **p. 14** *tl, cr,* **p. 19** *tl, cr,* **p. 24** *tl, cr,* **p. 32** *tl, cr,* **p. 36** *tl, cr,* **p. 42** *tl, cr,* **p. 46** *tl, cr,* **p. 53** *tl, cr,* **p. 58** *tl, cr,* **p. 62** *tl, cr,* **p. 65** *tl, cr,* **p. 69** *tl, cr,* **p. 75** *tl, cr,* **p. 82** *tl, cr,* **p. 89** *tl, cr,* **p. 92** *tl, cr,* **p. 96** *tl, cl* © Stocker Team/Adobe Stock.

t = top, *b* = bottom, *l* = left, *r* = right, *c* = centre

Every effort has been made to trace all copyright holders, but if any have been inadvertently overlooked, the Publishers will be pleased to make the necessary arrangements at the first opportunity.

Hachette UK's policy is to use papers that are natural, renewable and recyclable products and made from wood grown in well-managed forests and other controlled sources. The logging and manufacturing processes are expected to conform to the environmental regulations of the country of origin.

Orders: please contact Hachette UK Distribution, Hely Hutchinson Centre, Milton Road, Didcot, Oxfordshire, OX11 7HH. Telephone: +44 (0)1235 827827. Email education@hachette.co.uk Lines are open from 9 a.m. to 5 p.m., Monday to Friday. You can also order through our website: www.hoddereducation.com

ISBN: 978 1 3983 0117 7

© Catherine Casey, Steph King and Josh Lury 2021
First published in 2017
This edition published in 2021 by
Hodder Education,
An Hachette UK Company
Carmelite House
50 Victoria Embankment
London EC4Y 0DZ
www.hoddereducation.com

Impression number   10 9 8 7 6
Year   2025 2024 2023

Cover illustration by Lisa Hunt, The Bright Agency

Illustrations by James Hearne, Natalie and Tamsin Hinrichsen, Vian Oelofsen

Typeset in FS Albert 17/19 by IO Publishing CC

Printed in Spain

A catalogue record for this title is available from the British Library.

Contents

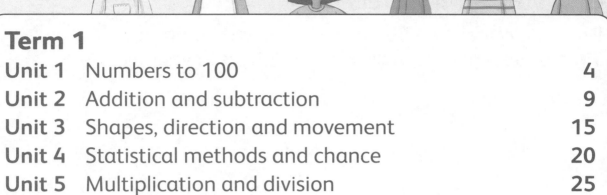

Unit 1   Numbers to 100

Remember: When you see this star ★, it is showing you that the activity develops your Thinking and Working Mathematically skills

Can you remember?

Draw a picture with 7 birds, 4 clouds and 3 houses.
Then draw 5 people, 2 cars and 6 flowers in your picture.

Reading and writing numbers to 20

1 Estimate how many there are. Then count to check.

		Estimate	Count
a	(cubes)		
b	(circles)		
c	(stars)		
d	(moons)		

2 Shade each number on the ten frames.

a 11

b 15

c 19

d 9

3 Write each number in words.

a 11 _____   b 17 _____

c 13 _____   d 19 _____

e 16 _____   f 14 _____

4 Draw 2 more patterns to make 10.

What do you notice?

Reading and writing numbers to 100

1 Fill in the missing numbers.

1	2	3	4	5	6	7		9	10
11		13	14	15	16	17	18	19	20
21	22	23	24		26	27	28	29	
31	32	33	34	35	36		38	39	40
41	42	43		45	46	47	48	49	50
	52	53	54	55		57	58	59	60
61	62	63	64	65	66	67	68	69	
71	72	73		75	76	77	78	79	80
81	82	83	84		86	87		89	90
91	92		94	95	96	97	98	99	100

2 Fill in the missing numbers.

a

| 20 | | | | | | | | | 30 |

b

| 90 | | | | | | | | | 100 |

3 Draw lines to match the words and numbers.

| thirty | forty-two | twenty | twenty-four | forty | thirty-one |

| 31 | 13 | 24 | 30 | 42 | 20 | 40 |

4 Choose 2 cards. Write as many numbers as you can.

| 9 | 1 | 0 | 1 |

_____

Estimating and counting to 100

1 How many ants?
Estimate. Then count to check.

Estimate: ____

Actual number: ____

2 Count on from each number in ones.

a 20, [21], [   ], [   ], [   ], [   ], [   ], [   ], [   ]

b 40, [41], [   ], [   ], [   ], [   ], [   ], [   ], [   ]

c 55, [56], [   ], [   ], [   ], [   ], [   ], [   ], [   ]

d 79, [   ], [   ], [   ], [   ], [   ], [   ], [   ], [   ]

3 Count down from fifty-five to thirty.

55, 54 _____

_____

_____

4 Complete the number grid by following the pattern.
Now check your answers by saying the numbers aloud.
Have you counted every number from 1 to 100?

				←	←				
			5	4	3	↑			
			6	1	2	↑			
			7	8	9	10			

Unit 1　Numbers to 100

Self-check

 I can do this.

 I can do this, but need to keep trying.

 I can't do this yet.

See how much you know!

What can I do?			
1　I can count up to 100 objects.			
2　I can say how many objects there are in a small group up to 10, without counting.			
3　I can make a good estimate of the number of objects in a group up to 100.			
4　I can count, saying the numbers in order, from 0 to 100.			
5　I can read numbers to 100.			
6　I can write numbers to 100.			

I need more help with:

_____

_____

_____

Can you remember?

Complete these:

4 + 5 = ☐   8 − 3 = ☐

5 + ☐ = 9   8 − ☐ = 3

10 + 4 = ☐   15 − 5 = ☐

4 + ☐ = 14   15 − ☐ = 11

The relationship between addition and subtraction

1 Draw the correct number of counters in the empty boxes.

a 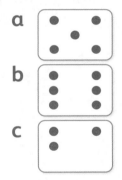 add 2 ☐ subtract 2 ☐

b 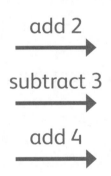 subtract 3 ☐ add 3 ☐

c 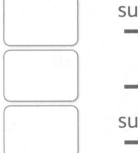 add 4 ☐ subtract 4 ☐

2 Write the addition and subtraction sentences for each problem.

a Maris has 3 apples. She eats 1. Then she picks 1 more apple.

☐ − ☐ = ☐ then ☐ + ☐ = ☐

b There are 4 birds in the tree. 2 more birds arrive.
Then 2 birds fly away.

☐ + ☐ = ☐ then ☐ − ☐ = ☐

c Jack has 7 cents. He spends 5 cents. Then he finds 5 cents.

☐ − ☐ = ☐ then ☐ + ☐ = ☐

Addition facts for 10 and related subtraction facts

1 Write the addition fact and the related subtraction fact each time.

a (6) + (4) = ( )   (10) − (4) = ( )

b ( ) + (5) = ( )   ( ) − (5) = ( )

c ( ) + ( ) = ( )   ( ) − ( ) = ( )

d ( ) + ( ) = ( )   ( ) − ( ) = ( )

2 Complete each number sentence.

a ( ) add 2 is 10, so 10 subtract 2 is ( )

b ( ) add 4 is 10, so ( ) subtract 4 is ( )

c ( ) and 3 is 10, so ( ) subtract ( ) is ( )

Pairs that total 20

1 Make up addition and subtraction sentences to match each bar model. Write as many as you can.

a
20	
9	11

b
20	
18	2

c
20	
7	13

_____   _____   _____

_____   _____   _____

_____   _____   _____

_____   _____   _____

2 Draw lines to join the pairs of numbers that make 20.

 12

 17

6

 15

 5

14

8

3

Addition and subtraction with numbers to 20

1 Complete each addition. Write the facts you used to help you.

12 + 5 = 17

Fact: **2 + 5 = 7**

a 11 + 6 = ⬜

Fact: _____

b 13 + 4 = ⬜

Fact: _____

c 13 + 5 = ⬜

Fact: _____

2 Complete each addition. Use 2 colours to show how you make 10 each time.

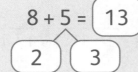

8 + 5 = 13

2   3

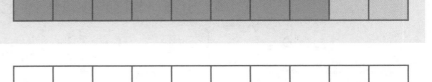

a 7 + 5 = ⬜

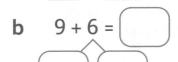

b 9 + 6 = ⬜

c 8 + 7 = ⬜

3 Complete each subtraction. Write the facts you used.

$18 - 7 =$ 11

Fact: $8 - 7 = 1$

a $18 - 6 =$ ☐

Fact: _____

b $17 - 4 =$ ☐

Fact: _____

c $17 - 5 =$ ☐

Fact: _____

4 Do each subtraction in 2 jumps. Jump 1 must land on 10.

a

0 1 2 3 4 5 6 7 8 9 10 11 12 13 14 15 16 17 18 19 20

$16 - 8 =$ ☐

b

0 1 2 3 4 5 6 7 8 9 10 11 12 13 14 15 16 17 18 19 20

$15 - 8 =$ ☐

c

0 1 2 3 4 5 6 7 8 9 10 11 12 13 14 15 16 17 18 19 20

$14 - 8 =$ ☐

d

0 1 2 3 4 5 6 7 8 9 10 11 12 13 14 15 16 17 18 19 20

$14 - 9 =$ ☐

Adding small numbers

1 The numbers in the shaded boxes should total 10.
Make each addition sentence correct.

a ☐ + ☐ + ☐ = 11

b ☐ + ☐ + ☐ = 14

c ☐ + ☐ + ☐ = 12

d ☐ + ☐ + ☐ = 15

2 The children are playing a game. Each child has 3 turns to throw a hoop over the cones.

Try to throw doubles or pairs that add up to 10.

a The table shows the children's scores. Work out the totals.

Name	1st hoop	2nd hoop	3rd hoop	Total
Viti	7	7	3	
David	5	5	7	
Zara	3	5	7	
Jack	3	3	3	

b Who has the lowest total score? _____

c Write the different totals you could get with 4 hoops.

_____

Adding multiples of 10

1 Draw a line from each flower to the correct pot to make 100.

Unit 2 Addition and subtraction

Self-check

 I can do this.

 I can do this, but need to keep trying.

 I can't do this yet.

See how much you know!

What can I do?	😄	😐	🙁
1 I can use models to show the inverse relationship between addition and subtraction.			
2 I can use complements of 10 to show the related addition and subtraction facts.			
3 I can identify complements of 20 from sets of numbers.			
4 I can add 2 numbers that total to 20.			
5 I can take away an amount from a number up to 20.			
6 I can estimate simple calculations to check the answers.			
7 I can use mental strategies to add small numbers.			
8 I can add pairs of multiples of 10, up to 100.			

I need more help with:

_____

_____

_____

Can you remember?

Write the name of each 3D shape.

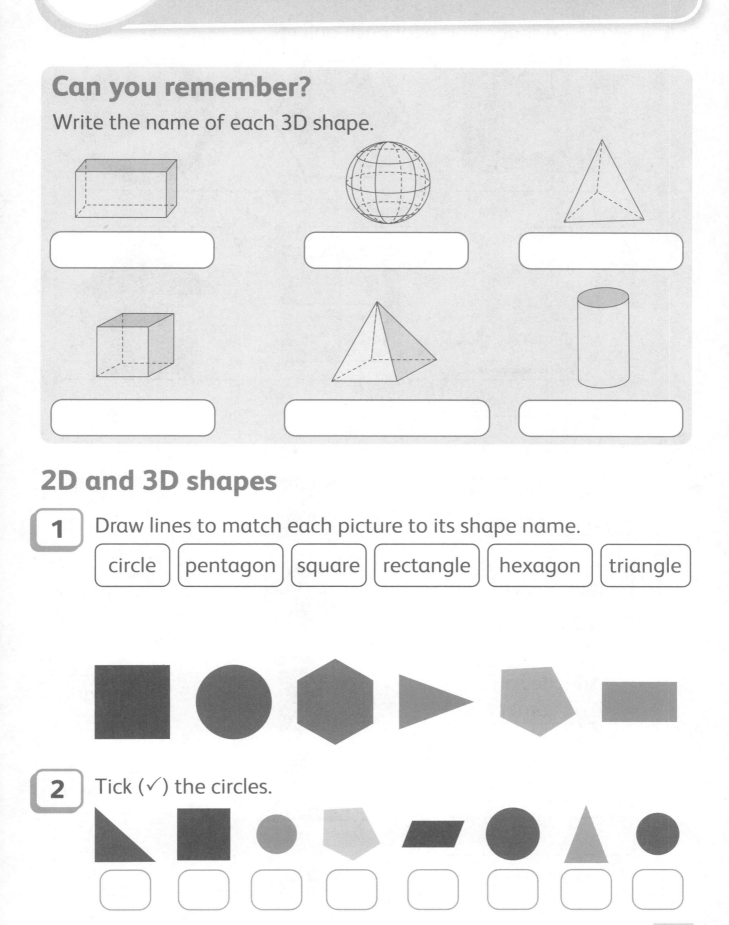

2D and 3D shapes

1 Draw lines to match each picture to its shape name.

| circle | pentagon | square | rectangle | hexagon | triangle |

2 Tick (✓) the circles.

3 Write the shape name for each object.

a

b

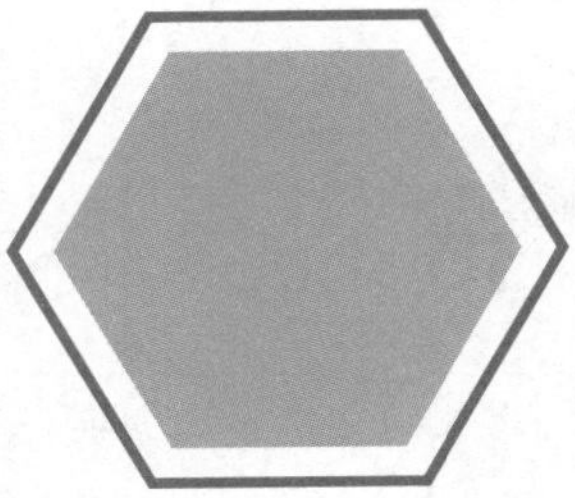

c

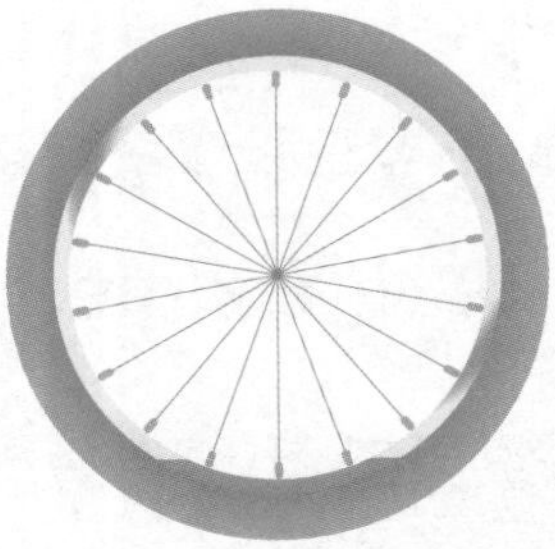

d

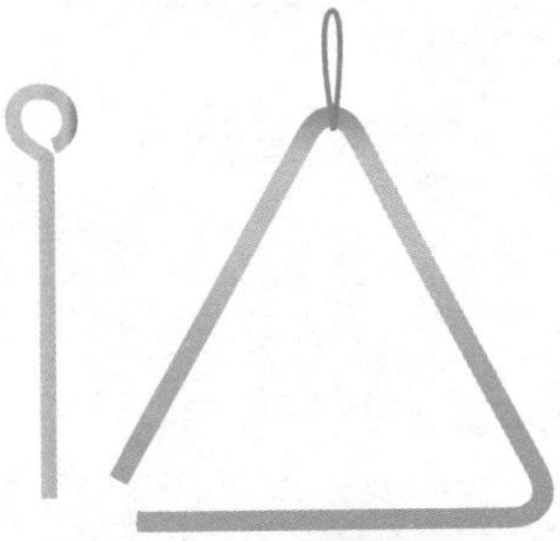

e

f

4 Draw lines to match each shape to its name and description.

three sides and three corners		pentagon
four sides and four corners		rectangle
six sides and six corners		square
five sides and five corners	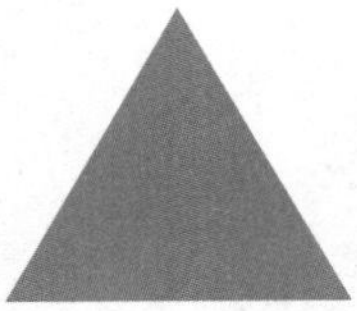	hexagon
four sides and four corners		triangle

5 Draw a line to match each description to its shape.

My shape has 6 square faces, 8 vertices and 12 edges.
What is my shape?

 a cube

 a cylinder

My shape has 1 square face, 4 triangular faces, 5 vertices and 8 edges.
What is my shape?

 a square-based pyramid

 a sphere

 6 Choose four 3D shapes with straight edges.
Complete the table.

Shape	Number of faces	Number of vertices	Number of edges

Patterns and pictures

1 Label the picture. Choose from these words.

sphere

cylinder

pyramid

cuboid

prism

cone

You need
pencil crayons

2 Follow these instructions to colour in the picture.

a Colour the spheres blue.   b Colour the pyramid purple.

c Colour the cylinder green.   d Colour the cone yellow.

e Colour the cuboid orange.   f Colour the prism red.

3 Draw the next 2 shapes in each pattern.

a

b

Unit 3 Shapes, direction and movement

Self-check

 I can do this.

 I can do this, but need to keep trying.

 I can't do this yet.

See how much you know!

What can I do?			
1 I can name some regular 2D shapes and some 3D shapes.			
2 I can recognise circles and explain key features.			
3 I can sort shapes in different ways and talk about their properties.			
4 I can compare shapes and say what is the same and what is different about them.			
5 I can use 2D shapes and 3D models to make patterns, pictures and models.			
6 I can use shapes to copy and continue sequences.			

I need more help with:

_____

_____

_____

Can you remember?

Draw a table to show how many of each there are.

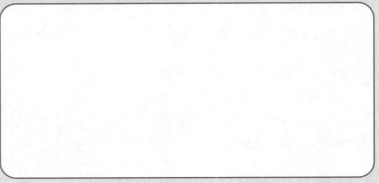

Pictograms and block graphs

1 An explorer recorded the number of animals he saw in a rainforest. He used the data to create a pictogram.

Rainforest animals		Key
Parrot	🔭🔭🔭🔭🔭🔭	🔭
Snake	🔭🔭🔭	= 1 animal
Tiger	🔭🔭🔭🔭🔭🔭🔭🔭🔭🔭🔭🔭🔭	
Monkey	🔭🔭🔭🔭🔭	
Lizard	🔭🔭	

Help Annay to complete the table by writing the total for each animal.

Animal	Tally	Number
Parrot	IIII I	6
Snake	III	
Tiger	IIII IIII I	
Monkey	IIII	
Lizard	II	

Which animal was most common?

_____

2 Jack asked his friends which shape they like best.
He recorded the data in a table. Help Jack to complete it.

Shape		Tally	Number of people
Star	★	卌 II	7
Circle	●	卌 IIII	
Square	■	III	
Triangle	▲	卌 III	

3 Use the data that Jack collected to draw a pictogram.

	Key

a Which shape is the most popular? _____

b Which shape is the least popular? _____

c How many children chose triangles? _____

d How many children chose stars? _____

4 Viti and Zara asked children how they travel to school. They started a block graph to show their data. Help them to complete it.

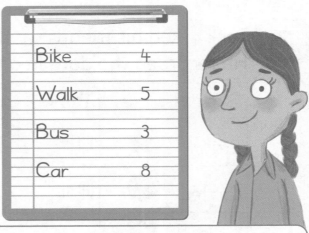

Bike   4

Walk   5

Bus   3

Car   8

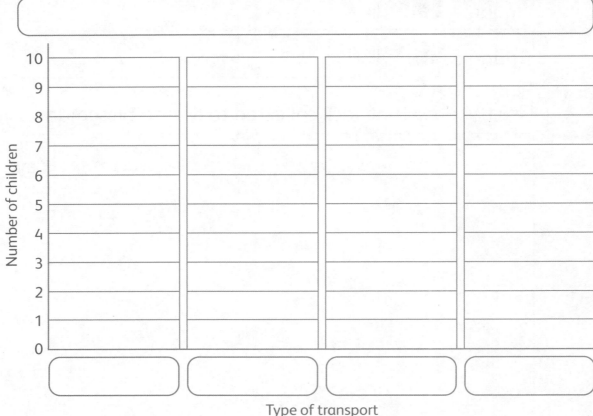

Number of children

Type of transport

a Which is the most popular way to travel to school? _____

b Which is the least popular way to travel to school? _____

c How many children travel by bus to school? _____

d How many children walk to school? _____

e How many more children walk than ride their bikes? _____

f How many children did Viti and Zara ask in total? _____

g How do you travel to school? _____

Venn diagrams and Carroll diagrams

 1 **a** Sort the numbers. Write them in the Venn diagram.

50   45   52   54   40   30   35   25   20   22   18

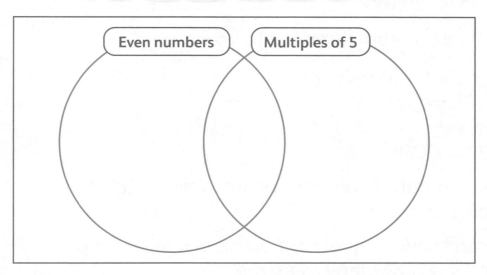

b Now add 6 more of your own numbers.

 2 Sort the shapes. Draw them to complete the Carroll diagram.

pentagon   semi-circle   trapezium   triangle   circle

square   rectangle   hexagon   oval   kite

	4 sides	Not 4 sides
Shaded		
Not shaded		

Self-check

 I can do this.

 I can do this, but need to keep trying.

 I can't do this yet.

See how much you know!

What can I do?			
1 I can organise information into a list, a table and a chart.			
2 I can show numbers of objects in a block graph and say how many there are.			
3 I can answer questions about a pictogram and a block graph.			
4 I can describe the data presented in tables, block graphs and pictograms.			
5 I can sort objects and shapes on a Carroll diagram, using 'not …' as a label.			
6 I can answer questions about Venn diagrams and Carroll diagrams.			

I need more help with:

_____

_____

_____

Can you remember?

Fill in the missing numbers.

Double 5 is ☐   Double ☐ is 6

Double 7 is ☐   Double ☐ is 18

Double 8 is ☐   Double ☐ is 20

Double 6 is ☐   Double ☐ is 8

Counting in twos, fives and tens

1   There are 10 fish in each group in the large tank.
How many fish are there in:

    a 2 groups? _____   **b** 3 groups? _____   **c** 4 groups? _____

2 Each hutch has 2 rabbits. How many rabbits are there in:

a 4 hutches? ☐

b 5 hutches? ☐

Remember to count in twos, fives or tens to solve the problems on this page.

c 6 hutches? ☐

d 7 hutches? ☐

3 Bird seed costs $5 for 1 box. Fish food costs $2 for 1 tin. What is the cost of:

a 5 tins of fish food?

b 8 boxes of bird seed?

c 3 tins of fish food and 3 boxes of bird seed?

d 6 tins of fish food and 6 boxes of bird seed?

e 5 tins of fish food and 3 boxes of bird seed?

4 Count back in twos. Write the numbers. Then complete each sentence.

a

		6	8	10

There are ☐ twos in 10.

b

						14

There are ☐ twos in 14.

c

									20

There are ☐ twos in 20.

Multiplication as repeated addition

1 Draw dots to show each repeated addition.
Complete the number sentences.

a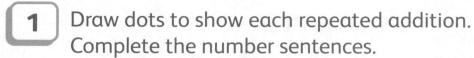

10 + 10 + 10 = ⬜   10 × 3 = ⬜

b ⬜ ⬜ ⬜ ⬜

5 + 5 + 5 + 5 = ⬜   5 × 4 = ⬜

c ⬜ ⬜ ⬜ ⬜

2 + 2 + 2 + 2 = ⬜   2 × 4 = ⬜

d ⬜ ⬜

10 + 10 = ⬜   10 × 2 = ⬜

2 Draw lines to match the calculations.

5 × 4	2 + 2 + 2 + 2 + 2
10 × 2	10 + 10 + 10 + 10 + 10
2 × 5	5 + 5 + 5 + 5 + 5 + 5
10 × 5	10 + 10
5 × 6	5 + 5 + 5 + 5

Using arrays to show multiplication

1 Write number sentences to match each array.

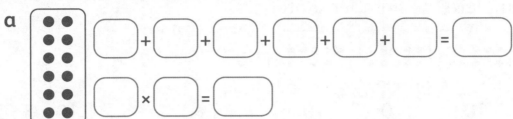

a ▢ + ▢ + ▢ + ▢ + ▢ + ▢ = ▢

▢ × ▢ = ▢

b ▢ + ▢ + ▢ = ▢

▢ × ▢ = ▢

c ▢ + ▢ + ▢ + ▢ + ▢ + ▢ + ▢ = ▢

▢ × ▢ = ▢

2 Write 2 multiplication sentences for each array.

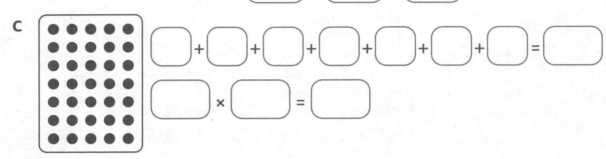

a	
b	
c	
d	

Division as sharing

1 Draw the counters each time. Write the division sentence to match.

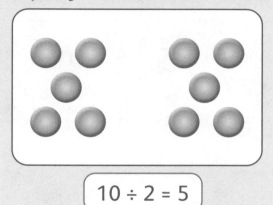

10 counters shared equally between 2

10 ÷ 2 = 5

a 16 counters shared equally between 2

◯ ÷ ◯ = ◯

b 20 counters shared equally between 10

◯ ÷ ◯ = ◯

c 35 counters shared equally between 5

◯ ÷ ◯ = ◯

2 Complete each division and bar model.

a 14 ÷ 2 = ◯

14

b 30 ÷ 5 = ◯

30

c 40 ÷ 10 = ◯

40

Division as grouping

Draw the objects and write a division fact.

 a Viti has 12 marbles. She makes groups of 2.
How many groups are there?

Division fact   12 ÷ 2 = ☐

b David has 20 marbles. He makes groups of 2.
How many groups are there?

Division fact ☐

2 Zara arranges her marbles in equal rows.

a How many rows of 10 marbles are there? ☐

b Complete the division fact.   ☐ ÷ 10 = ☐

Division as repeated subtraction

1 The children take away blocks until there are none left.

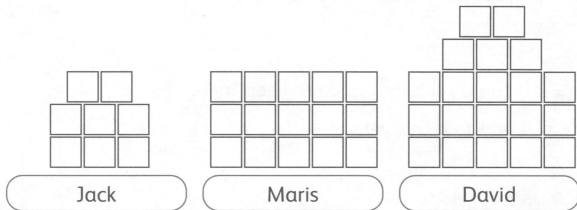

Jack   Maris   David

a Jack takes away 2 blocks at a time.
How many times can he do this? ◻

b Maris takes away 5 blocks at a time.
How many times can she do this? ◻

c David takes away 10 blocks at a time.
How many times can he do this? ◻

2 Draw jumps on the number lines to match each division.

a 15 ÷ 5 = ◻

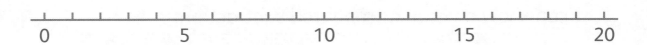

b 50 ÷ 10 = ◻

c 18 ÷ 2 = ◻

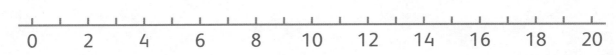

31

Unit 5　Multiplication and division

Self-check

 I can do this.

 I can do this, but need to keep trying.

 I can't do this yet.

See how much you know!

What can I do?			
1　I can count in steps of 2, 5 and 10.			
2　I can show multiplication as repeated addition.			
3　I know how to write number sentences for multiplication using the symbols × and =.			
4　I can use arrays to show multiplication.			
5　I know how to write number sentences for division using the symbols ÷ and =.			
6　I can use sharing to work out a division.			
7　I can use grouping to work out a division.			
8　I can show division as repeated subtraction.			

I need more help with:

_____

_____

_____

Can you remember?

	months in a year		days in a week
	minutes in an hour		seconds in a minute

Time

1 Choose the best unit for measuring the time of each event. Draw lines to match.

brushing your teeth

a plane journey

winking your eye

seconds

minutes

hours

2 **a** How many days are there in 5 weeks? _____

      b How many months are there in 2 years? _____

3 Match the lengths of time. One has been done for you.

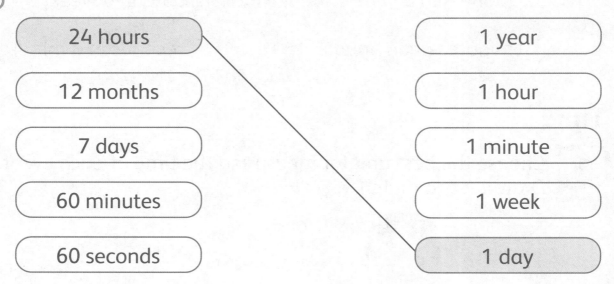

24 hours	1 year
12 months	1 hour
7 days	1 minute
60 minutes	1 week
60 seconds	1 day

4 Draw the hands on each clock to show the time.

a

3:35

b

10:45

c

6:20

d

8:10

Length

1 Which objects are longer than 1 metre? Circle them.

2 Which objects are shorter than 1 metre? Circle them.

3 Draw something that is longer than 1 metre.

4 Draw something that is shorter than 1 metre.

Self-check

 I can do this.

 I can do this, but need to keep trying.

 I can't do this yet.

See how much you know!

What can I do?			
1 I can use and compare units of time.			
2 I can read and write the time to the nearest 5 minutes.			
3 I can use non-standard units to measure the lengths of objects.			
4 I can use a metre rule to measure the lengths of objects.			
5 I can find out if something is longer or shorter than 1 metre.			
6 I can estimate the lengths of objects in whole metres before measuring.			

I need more help with:

_____

_____

_____

Can you remember?

Write 5 different ways to total 10.

☐ + ☐ = 10   ☐ + ☐ = 10   ☐ + ☐ = 10

☐ + ☐ = 10   ☐ + ☐ = 10

Addition and subtraction facts for 20

1 Make **20** in different ways. Fill in the missing numbers.

a   20 → 11, ○   11 + ☐ = 20

b   20 → 13, ○   13 + ☐ = 20

c   20 → 14, ○   14 + ☐ = 20

d   20 → 5, ○   5 + ☐ = 20

2 Write the inverse number sentences.

a   4 + 16 = 20 _____   b   20 − 2 = 18 _____

c   12 + 8 = 20 _____   d   20 − 6 = 14 _____

3 a   A pie takes 20 minutes to bake.
It has been baking for 14 minutes.
How many minutes are left?   ☐ minutes

b   There are 20 pencils in a pack.
Maris uses 3 pencils. How many are left?   ☐ pencils

c   11 boys and 9 girls are on the bus.
How many children are there in total?   ☐ children

Adding and subtracting multiples of 10

1 Fill in the missing numbers.

a
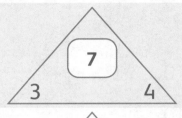

$3 + 4 = \boxed{}$ $\boxed{} - 3 = 4$

$4 + 3 = \boxed{}$ $\boxed{} - 4 = 3$

b
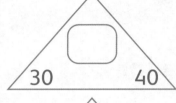

$30 + 40 = \boxed{}$ $\boxed{} - 30 = 40$

$40 + 30 = \boxed{}$ $\boxed{} - 40 = 30$

c
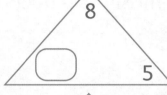

$\boxed{} + 5 = 8$ $8 - 5 = \boxed{}$

$5 + \boxed{} = 8$ $8 - \boxed{} = 5$

d
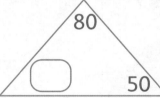

$\boxed{} + 50 = 80$ $80 - 50 = \boxed{}$

$50 + \boxed{} = 80$ $80 - \boxed{} = 50$

What do you notice?

2 Draw more 10 cent coins to make the correct total.
Write the addition sentences.

a
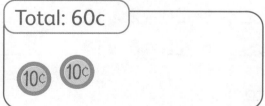

b
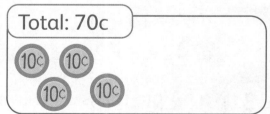

c
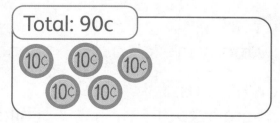

d
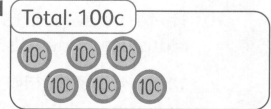

3 Write an addition and a subtraction fact for each bar model.

a ( ) + ( ) = ( )

100	
10	90

( ) – ( ) = ( )

b ( ) + ( ) = ( )

90	
70	20

( ) – ( ) = ( )

c ( ) + ( ) = ( )

80	
50	30

( ) – ( ) = ( )

d ( ) + ( ) = ( )

60	
20	40

( ) – ( ) = ( )

Making estimates

1 Draw lines to match the calculations with the estimates.

| answer will be less than 10 | answer will be more than 10 |

(14 – 3) (4 + 5) (18 – 6) (16 – 8) (3 + 8)

Adding small numbers

1 Colour in the 2 numbers you will add first. Then find the total of all 3 or 4 numbers. Share your ideas with a friend.

a (5) + (3) + (5) = ( )   b (1) + (3) + (7) + (4) = ( )

c (4) + (4) + (6) = ( )   d (5) + (3) + (3) + (5) = ( )

2 The arrows score the number they land on.
Find the total score each time.

a 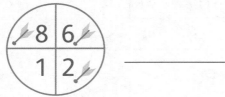 _____

b  _____

3 Annay buys 3 lemons, 5 apples, 7 bananas and 5 pears.
How many fruits does he buy in total?

Adding two-digit and one-digit numbers

1 Show jumps on the number line to complete these additions.

a 23 + 4 = ⬭

20 21 22 23 24 25 26 27 28 29 30

b 33 + 4 = ⬭

30 31 32 33 34 35 36 37 38 39 40

c 43 + 3 = ⬭

40 41 42 43 44 45 46 47 48 49 50

2 Draw ones and sticks of tens to add the numbers.
Use 2 different colours.

a 36 + 2 = ⬭   b 25 + 3 = ⬭

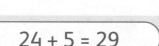

24 + 5 = 29

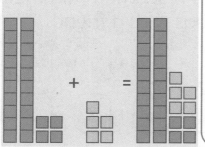

Subtracting a one-digit number from a two-digit number

Remember to subtract in the order the numbers are given.

1 Draw the jumps on the number lines. Complete the number sentences.

a 24 – 3 = ☐

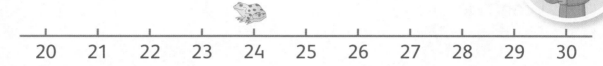

| 20 | 21 | 22 | 23 | 24 | 25 | 26 | 27 | 28 | 29 | 30 |

b 37 – 5 = ☐

| 30 | 31 | 32 | 33 | 34 | 35 | 36 | 37 | 38 | 39 | 40 |

c 48 – 2 = ☐

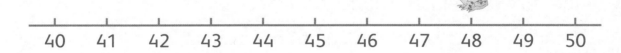

| 40 | 41 | 42 | 43 | 44 | 45 | 46 | 47 | 48 | 49 | 50 |

d 56 – 3 = ☐

| 50 | 51 | 52 | 53 | 54 | 55 | 56 | 57 | 58 | 59 | 60 |

2 Jack has 38 marbles. He gives 6 to his friend. How many does he have left? ☐

3 Zara has 45 pencil crayons. She loses 3 of them. How many does she have left? ☐

4 Annay thinks of a number. He takes away 4. He has 23 left. What number did Annay start with? ☐

Unit 7 Addition and subtraction

Self-check

 I can do this.

 I can do this, but need to keep trying.

 I can't do this yet.

See how much you know!

What can I do?			
1 I know complements of 20 and can identify related subtraction facts.			
2 I can show the inverse subtractions for an addition sentence and the inverse additions for a subtraction sentence.			
3 I can add pairs of multiples of 10 up to 100 and identify the related subtraction facts.			
4 I can estimate simple calculations to check the answers.			
5 I can use mental strategies to add small numbers.			
6 I can add any number up to 9 to a 2-digit number, without crossing a tens boundary.			
7 I can take away any number up to 9 from a 2-digit number, with no regrouping.			

I need more help with:

_____

_____

_____

Can you remember?

Draw any coins and notes that you can use at a pretend shop.

Understanding coins and notes

1 Draw lines to match each coin and note to its value.

| 1 dollar | 10 cents | 25 cents | 20 dollars |

| 1 cent | 5 dollars | 5 cents | 10 dollars |

2 Draw each set of money in order of value.

a

b

c

3 Draw 1c coins to the same value as the coins in each hand.

a

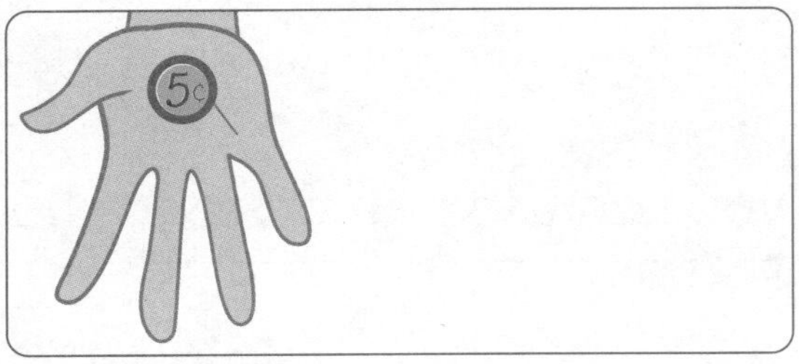

b

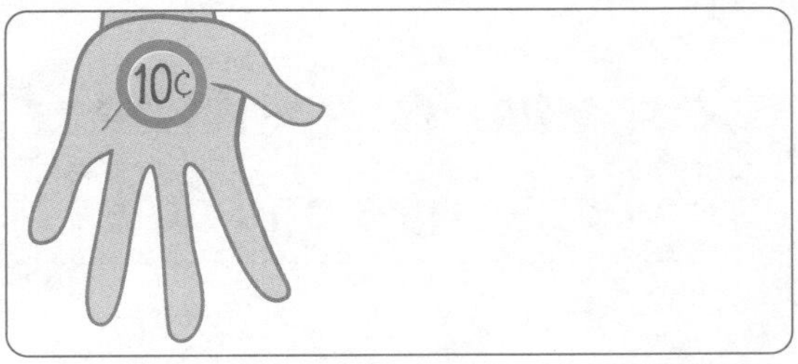

c

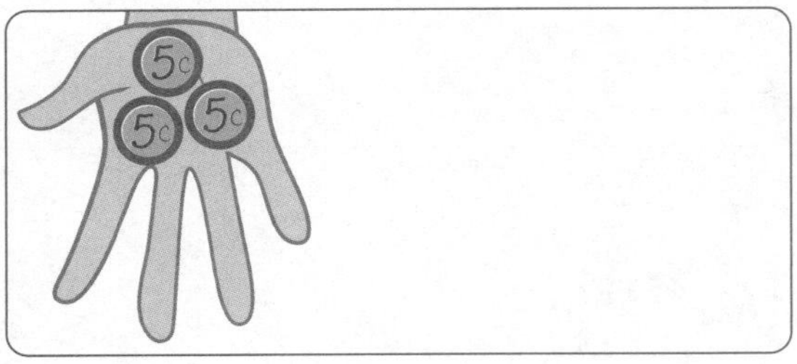

d

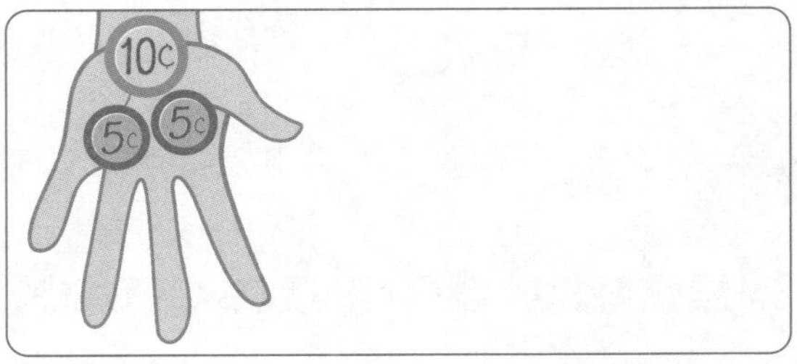

4 Write about something you would like to buy for a friend.
Would you buy the gift with coins or notes?

_____

_____

Working out the total amount

1 Complete each part-part-whole model.

a

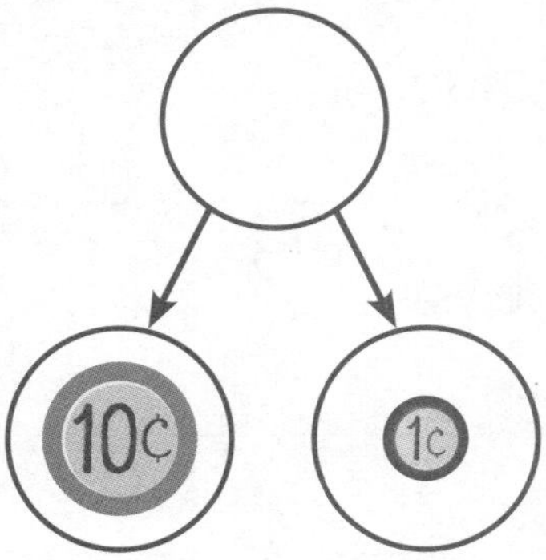

10¢   1¢

b

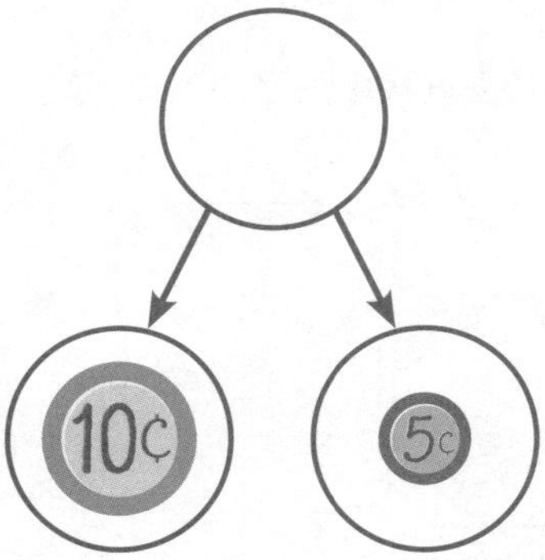

10¢   5¢

c

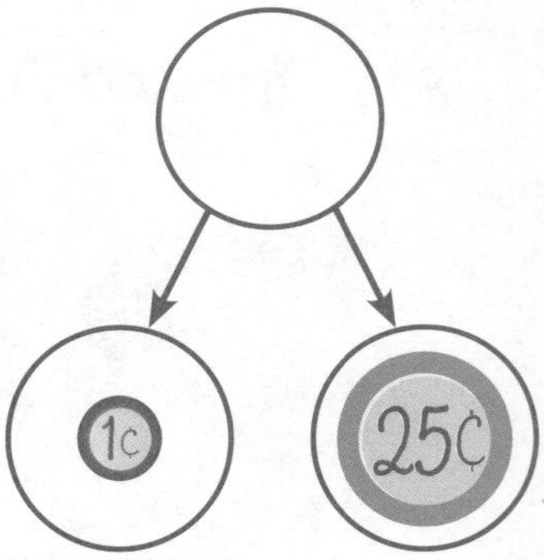

1¢   25¢

d

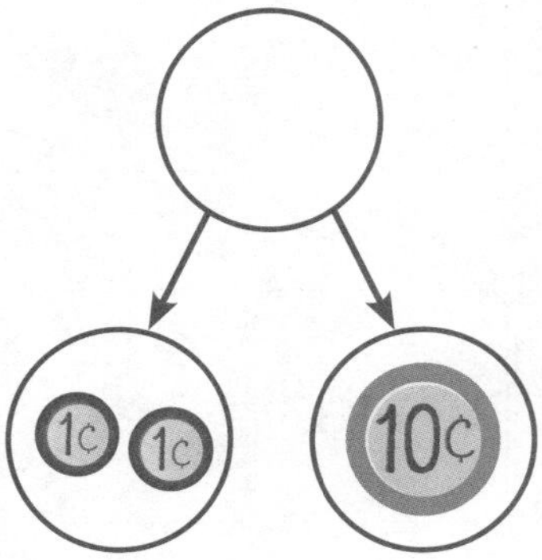

1¢ 1¢   10¢

e

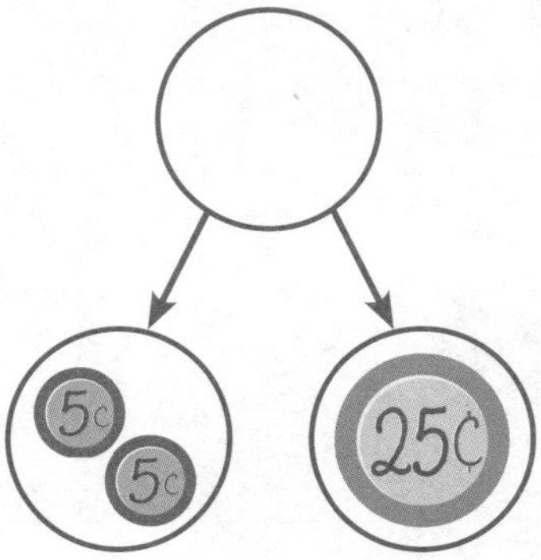

5¢ 5¢   25¢

f

1¢ 5¢   10¢

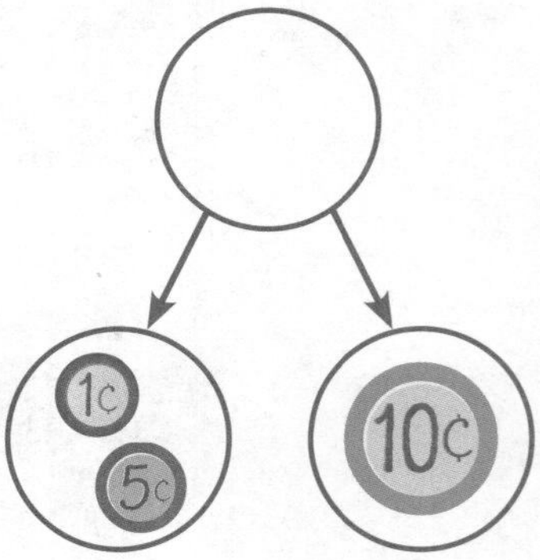

Making amounts

1 Match each purse or hand to the correct amount of money.

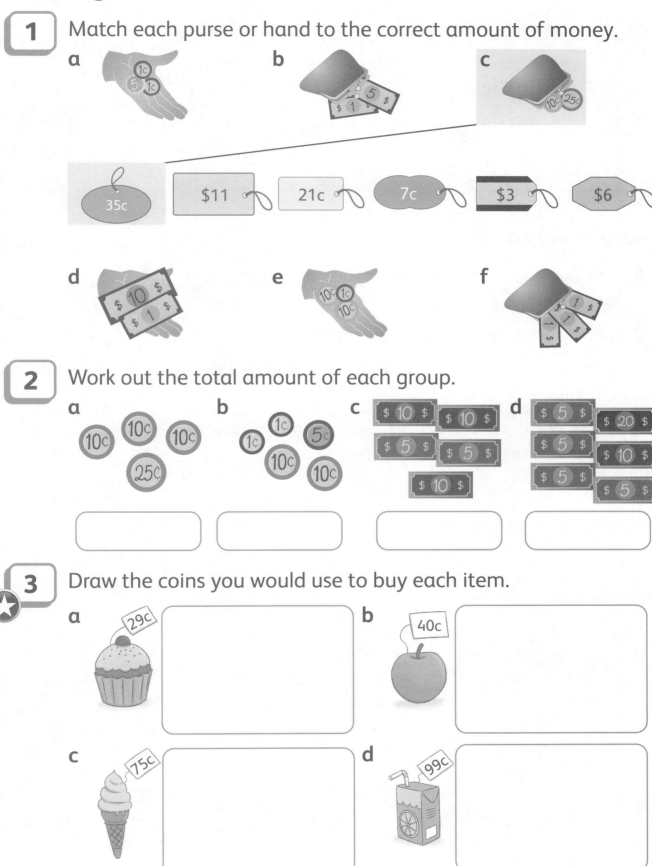

2 Work out the total amount of each group.

3 Draw the coins you would use to buy each item.

Unit 8   Money

Self-check

 I can do this.

 I can do this, but need to keep trying.

 I can't do this yet.

See how much you know!

What can I do?			
1 I know the coins and notes that are used as currency.			
2 I can sort coins and make sets of different values.			
3 I can compare values and put coins in order of value.			
4 I can combine coins and notes to make different values to buy different items.			

I need more help with:

_____

_____

_____

Can you remember?

Estimate the number and then count to check.

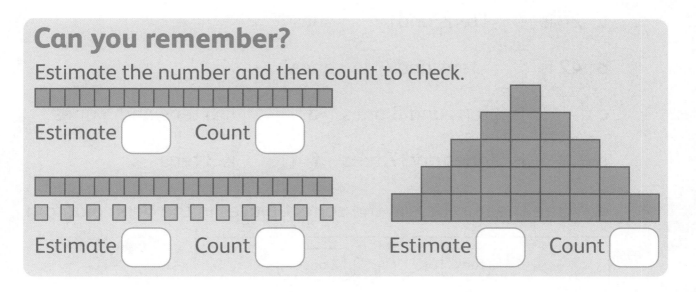

Estimate ☐   Count ☐

Estimate ☐   Count ☐

Estimate ☐   Count ☐

Tens and ones

1   Write each number as tens and ones.

a

10s	1s

b

10s	1s

c

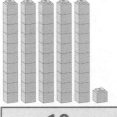

10s	1s

d

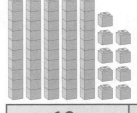

10s	1s

e

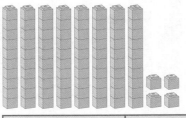

10s	1s

f

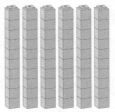

10s	1s

2 Complete each number sentence to show the place value.

a 24 is ☐ tens and ☐ ones

b 42 is ☐ tens and ☐ ones

c ☐ is 8 tens and 8 ones   **d** ☐ is 6 tens and 5 ones

e ☐ is 5 ones and 7 tens   **f** ☐ is 3 tens

3 **a** Write the numbers in the correct places in the Venn diagram.

2 tens   2 ones

42
22
52
21   32
23   25
26

b Write 5 or more of your own numbers in the diagram.

Counting in tens and ones

1 **a** How many cubes each time? What patterns do you notice?

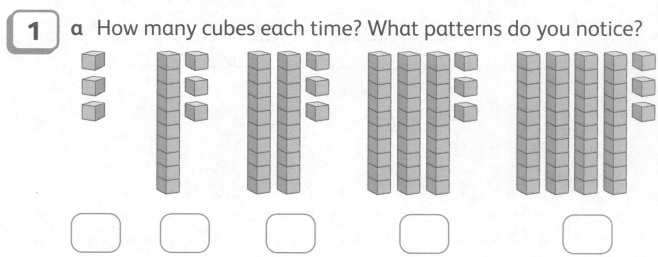

☐   ☐   ☐   ☐   ☐

Unit 9 Number patterns and place value

b Draw a picture to show the counting pattern: 8, 18, 28, 38, 48.

2

a Make your own place value apparatus.

You need	Method
• wooden lolly sticks	**Make tens:** Count out 10 lolly sticks. Tie them together with string. Repeat 4 times.
• string, ribbon or elastic bands	**Make ones:** Use single lolly sticks.

b Now make these numbers. Draw lolly sticks in tens and ones.

13	14	15

23	24	25

43	44	45

Counting in fives

1 Start on 5 on this 100 grid. Count aloud in fives.
Colour in all the numbers you say.

1	2	3	4	5	6	7	8	9	10
11	12	13	14	15	16	17	18	19	20
21	22	23	24	25	26	27	28	29	30
31	32	33	34	35	36	37	38	39	40
41	42	43	44	45	46	47	48	49	50
51	52	53	54	55	56	57	58	59	60
61	62	63	64	65	66	67	68	69	70
71	72	73	74	75	76	77	78	79	80
81	82	83	84	85	86	87	88	89	90
91	92	93	94	95	96	97	98	99	100

2 How many cubes? Count in fives.

a cubes

b cubes

c cubes

d cubes

3 Circle the dots
in groups of 5
to count them.

There are ☐

dots in total.

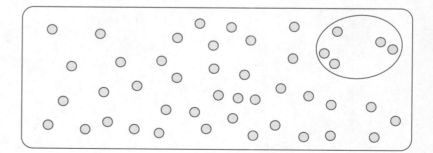

4 Continue counting each time.

a 1, 6, 11, 16, ⬚ , ⬚ , ⬚ , ⬚ , ⬚ , ⬚ , ⬚ , ⬚

b 8, 13, 18, ⬚ , ⬚ , ⬚ , ⬚ , ⬚ , ⬚ , ⬚ , ⬚

c 37, 42, 47, ⬚ , ⬚ , ⬚ , ⬚ , ⬚ , ⬚ , ⬚ , ⬚

d 70, 65, 60, ⬚ , ⬚ , ⬚ , ⬚ , ⬚ , ⬚ , ⬚ , ⬚

e 88, 83, 78, ⬚ , ⬚ , ⬚ , ⬚ , ⬚ , ⬚ , ⬚ , ⬚

f 99, 94, 89, ⬚ , ⬚ , ⬚ , ⬚ , ⬚ , ⬚ , ⬚ , ⬚

Counting in twos

1 Continue counting and making circles on the number line.

2 Circle the even numbers. Colour in the odd numbers.
Now count back in twos from 40.

1	2	3	4	5	6	7	8	9	10
11	12	13	14	15	16	17	18	19	20
21	22	23	24	25	26	27	28	29	30
31	32	33	34	35	36	37	38	39	40

3 Sort each number into the table.

64, 65, 66, 67, 68, 69, 70, 71, 72, 73, 74, 75

Odd	Even

4 Tick the bags of sweets that you can share into 2 equal groups.

a

b

c

d

e

f

Self-check

See how much you know!

 I can do this.

 I can do this, but need to keep trying.

 I can't do this yet.

What can I do?			
1 I can count on in ones, twos, fives and tens from any number up to 100.			
2 I can count back in ones, twos, fives and tens from any number on a 100 grid.			
3 I can recognise that an even number of objects can be shared into 2 equal groups and odd numbers cannot.			
4 I can read and write 2-digit numbers and show what each digit stands for.			
5 I understand that zero (0) acts as a placeholder in a number.			

I need more help with:

_____

_____

_____

Can you remember?

How heavy is each item? Read the scales.

a   b   c

Time

1 Draw lines to match the times to the digital clocks.

twenty past 9

quarter past 11

five past 3

50 minutes past 8

half-past 6

2 Draw hands on the analogue clocks to match the time on each digital clock.

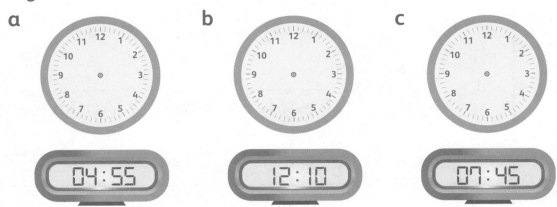

a   b   c

04:55   12:10   07:45

3 Write the time on each digital clock.

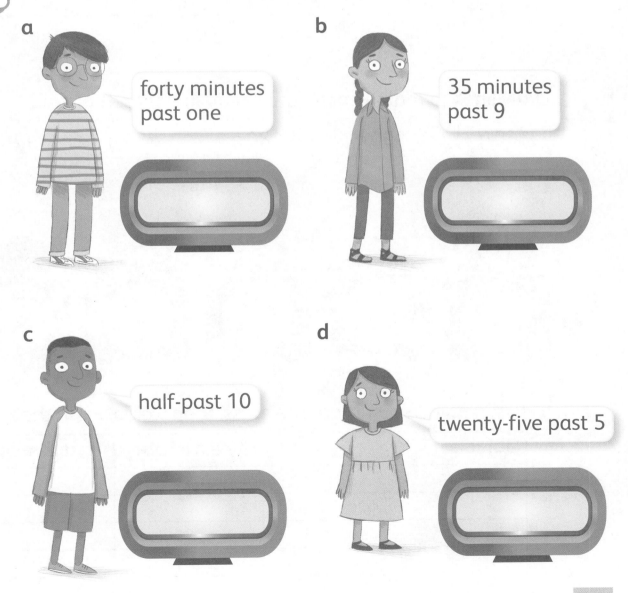

a

forty minutes past one

b

35 minutes past 9

c

half-past 10

d

twenty-five past 5

Length

1 Estimate and then measure each line.

a   b   c ───────────   d

e

Line	My estimate	Actual measure
a		
b		
c		
d		
e		

Mass

1 Draw lines to match each parcel to its mass on the scale.

160 g   190 g   180 g   250 g   170 g

2 **a** Pick 4 objects. Estimate the mass, then weigh. Fill in the table.

Object	My estimate	Actual mass

b Draw your objects in order from lightest to heaviest.

Lightest			Heaviest

3 **a** Which items are heavier than 1 kg? Make a tick (✓) in the box.

b Which items are lighter than 1 kg? Make a cross (✗).

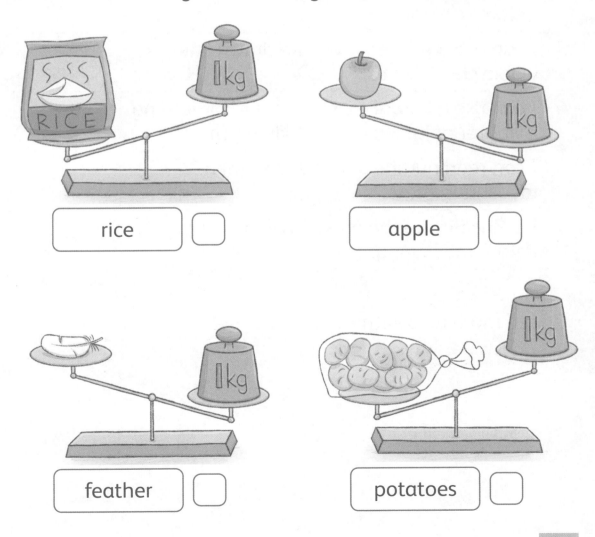

rice ☐

apple ☐

feather ☐

potatoes ☐

Self-check

 I can do this.

 I can do this, but need to keep trying.

 I can't do this yet.

See how much you know!

What can I do?			
1 I can read and write times shown to the nearest 5 minutes.			
2 I can estimate lengths in whole centimetres before measuring.			
3 I can use a ruler to measure drawn lines in whole centimetres.			
4 I can use a balance to find out if something is lighter or heavier than 1 kilogram.			
5 I can read numbers on a scale when measuring mass.			
6 I can estimate the mass of an object before measuring.			

I need more help with:

_____

_____

_____

Can you remember?

Draw a line of symmetry on each shape.

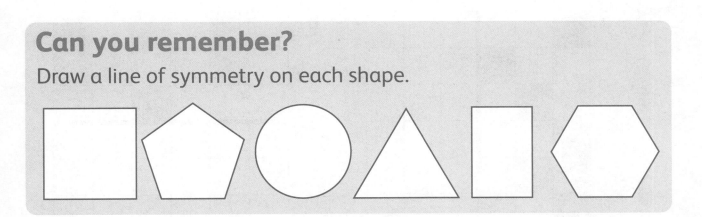

Symmetry

1 Tick (✓) each shape that has a line of symmetry.

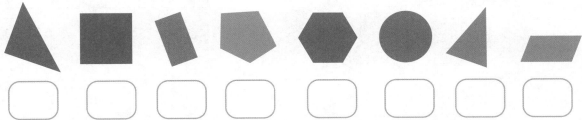

2 a Draw a line of symmetry on the shapes in each question.

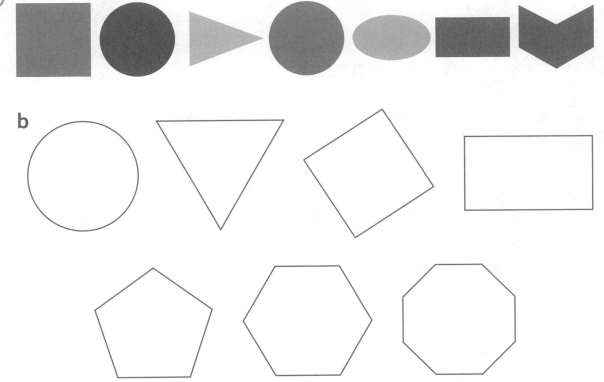

b

3 Complete the patterns to make them symmetrical.

a
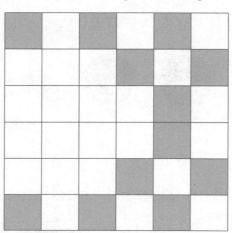

b
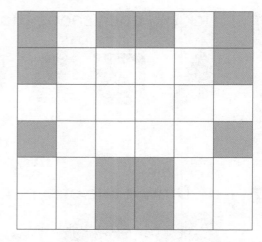

4 Draw a line of symmetry on each pattern.

a

b

Position and movement

1 How many times does each shape look the same through one whole turn of the shape?

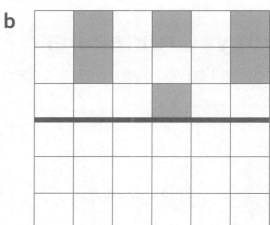

a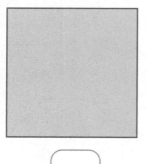

b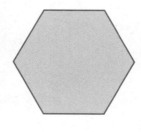

c

2 The robot can move forwards, backwards, and turn clockwise and anticlockwise.

Imagine that you are a robot. Which way will you go?

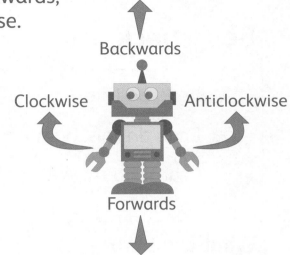

Backwards

Clockwise   Anticlockwise

Forwards

Help the robot to move through the maze. Follow each set of directions. Which shape does the robot walk to each time?

a Forwards 3 squares
A quarter turn anticlockwise
Forwards 3 squares
A quarter turn clockwise
Backwards 2 squares

Shape: _____

b Forwards 3 squares
A quarter turn anticlockwise
Forwards 3 squares
A quarter turn clockwise
4 squares forwards
A quarter turn anticlockwise
Forwards 4 squares

Shape: _____

Can you give instructions for the robot to reach another shape?

Unit 11　Shapes, direction and movement

Self-check

 I can do this.

 I can do this, but need to keep trying.

 I can't do this yet.

See how much you know!

What can I do?			
1 I can complete a symmetrical picture by drawing the other half.			
2 I can draw a line of symmetry on a 2D shape.			
3 I can work out how many times a shape looks the same through a whole turn of the shape.			
4 I can follow and give instructions to make clockwise and anticlockwise turns.			

I need more help with:

_____

_____

_____

Unit 12 Fractions

Can you remember?

Show how to break up these shapes into halves.

  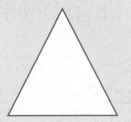

Equal parts

1 Shade one quarter of each shape.

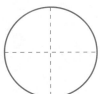  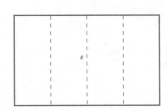

2 Divide each shape into quarters in different ways.

    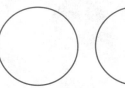 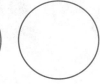

3 Tick (✓) the shapes that show quarters.

a 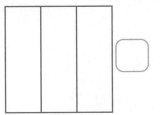   b 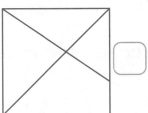   c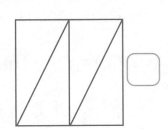

d 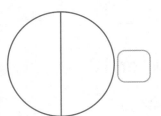   e 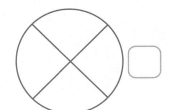   f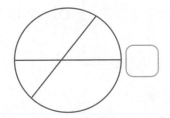

63

4 Explain the difference between halves ▯ and quarters ▯.

a Halves _____

_____

b Quarters _____

_____

Fractions of a group

1 Colour in one quarter of each set. Write how many of each.

a

$\frac{1}{4}$ of ☐ 8 ☐ = ☐

b

$\frac{1}{4}$ of ☐ = ☐

2

One quarter of 20 stars is 4 stars.

Draw a picture to show that Jack is wrong.

3 Complete these fraction sentences.

a $\frac{1}{2}$ of 8 is ☐

$\frac{1}{4}$ of 8 is ☐

b $\frac{1}{4}$ of 4 is ☐

$\frac{1}{2}$ of ☐ is 4

c $\frac{1}{2}$ of 16 is ☐

$\frac{1}{4}$ of 16 is ☐

d $\frac{1}{2}$ of 10 is ☐

$\frac{1}{4}$ of ☐ is 10

Unit 12   Fractions

Self-check

 I can do this.

 I can do this, but need to keep trying.

 I can't do this yet.

See how much you know!

What can I do?	😊	😐	😟
1 I can fold shapes to show quarters.			
2 I can find one quarter of a shape.			
3 I know that one quarter is 1 of 4 equal parts.			
4 I can find $\frac{1}{4}$ of a set or group.			
5 I know that the word 'fraction' is equal parts of a whole.			

I need more help with:

_____

_____

_____

Unit 13 Statistical methods and chance

Can you remember?

Is there a pattern?   Yes ☐   No ☐

Pictograms and block graphs

1   **a**   Complete this table.

Day of the week	Tally	Drinks Viti sold
Monday	IIII	
Tuesday		7
Wednesday		3
Thursday		6
Friday	IIII IIII	

b   Use the data in the table to complete the block graph below.

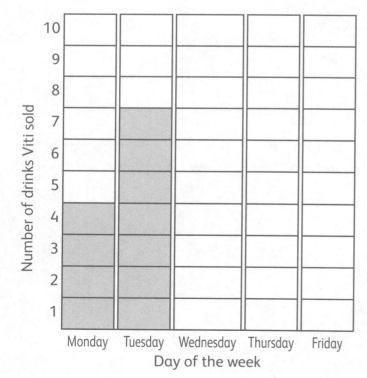

c   How many drinks did Viti sell on Thursday? ☐

d   On which day did Viti sell the most drinks?

_____

e   On which day did Viti sell 7 drinks?

_____

f   How many drinks did Viti sell on Thursday and Friday altogether? ☐

Venn diagrams and Carroll diagrams

 1 Draw the shapes in the correct places on the Venn diagram.

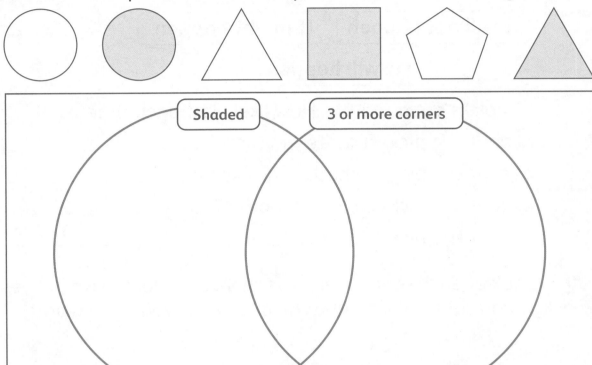

 2 Draw the shapes in the correct boxes of the Carroll diagram.

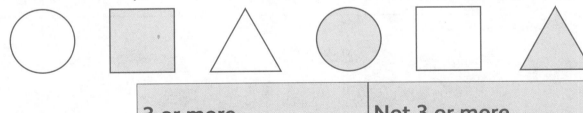

	3 or more corners	Not 3 or more corners
Shaded		
Not shaded		

Chance

1 Use these statements to answer the questions below.

| It will not happen | It might happen |

| It will happen |

Viti picks out a marble. How likely is it each time:

a that she picks a white marble? _____

b that she picks a shaded marble? _____

c that she picks a striped marble? _____

d that she picks a marble? _____

2 Look at each set of shapes. Is there a pattern? Tick (✓) yes or no. If you said yes, describe the pattern in your own words.

a

Yes ☐ No ☐ _____

_____

b

Yes ☐ No ☐ _____

_____

c

Yes ☐ No ☐ _____

_____

Unit 13 Statistical methods and chance

Self-check

 I can do this.

 I can do this, but need to keep trying.

I can't do this yet.

See how much you know!

What can I do?	😉	😐	🙁
1 I can organise information into a list, a table and a chart.			
2 I can show numbers of objects in a block graph and say how many there are.			
3 I can answer questions about a pictogram and a block graph.			
4 I can sort objects and shapes on a Carroll diagram, using 'not …' as a label.			
5 I can answer questions about Venn diagrams and Carroll diagrams.			
6 I can describe the data presented in tables, block graphs and pictograms.			
7 I can recognise and describe patterns and non-patterns.			
8 I can say how likely something with more than one outcome is.			

I need more help with:

_____

_____

Can you remember?

What number is shown each time?

Patterns and ordinals

1 Look at the picture and answer the questions.

a What is 1st? _____

b What is 11th? _____

c What is 15th? _____

d What positions are the stars? _____

2 Continue the pattern.

a What shape is 21st?

_____

b What shape is 24th?

_____

c What shape is 29th?

_____

1	2	3	4	5	6
7	8	9	10	11	12
13	14	15	16	17	18
19	20	21	22	23	24
25	26	27	28	29	30

3 Make your own repeating pattern. Use circles, triangles and squares. The 11th shape should be a circle. The 12th shape should be a triangle. The 13th shape should be a square.

Tens and ones

1 Complete each part-part-whole model.

a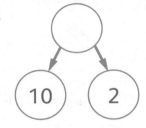

b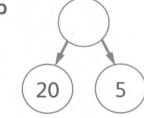

c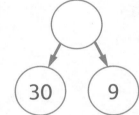

d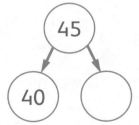

e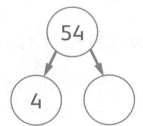

f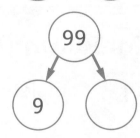

2 Draw lines to match the numbers to the addition sentences.

| 20 + 4 | 40 + 2 | 3 + 30 | 3 + 40 |

| 42 | 34 | 55 | 24 | 15 | 43 | 33 | 51 |

| 30 + 4 | 50 + 1 | 10 + 5 | 5 + 50 |

3 Show different ways to regroup the number 55.
Can you see any patterns in your numbers?

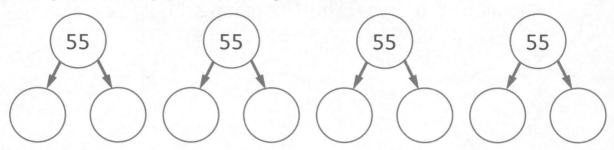

4 Complete these sets of number sentences.

a	b	c	d
20 + 1 = [ ]	[ ] = 10 + 1	[ ] = 30 + 4	70 + [ ] = 74
20 + 2 = [ ]	[ ] = 30 + 1	[ ] = 40 + 3	[ ] + 4 = 94
20 + 3 = [ ]	[ ] = 50 + 1	54 = 4 + [ ]	60 + [ ] = 74
[ ] = 20 + 4	1 + 60 = [ ]	50 + [ ] = 55	44 = 14 + [ ]

Comparing and ordering numbers

1 Complete each statement using the following words:

greater than	less than

a 13 is _____ 14   b 15 is _____ 50

c 14 is _____ 13   d 51 is _____ 50

e 31 is _____ 41   f 51 is _____ 60

g 41 is _____ 31   h 61 is _____ 51

2 Use the number lines to help you order each set of numbers.

a
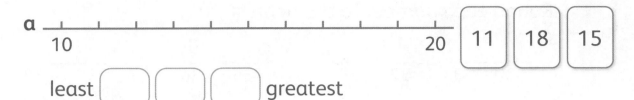

least ☐ ☐ ☐ greatest

b
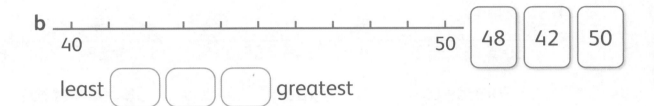

least ☐ ☐ ☐ greatest

c
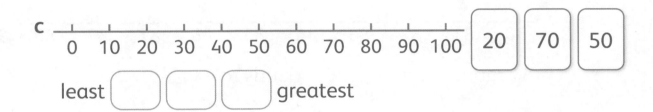

least ☐ ☐ ☐ greatest

d
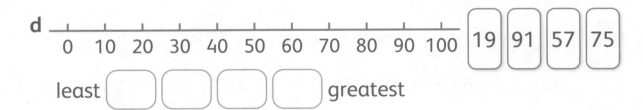

least ☐ ☐ ☐ ☐ greatest

e
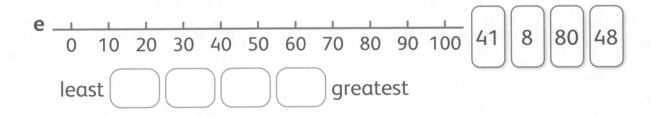

least ☐ ☐ ☐ ☐ greatest

3 Mark these numbers on the number line. 75, 12, 90, 51, 4, 40

Round to the nearest 10

1 Round each number to the nearest 10.

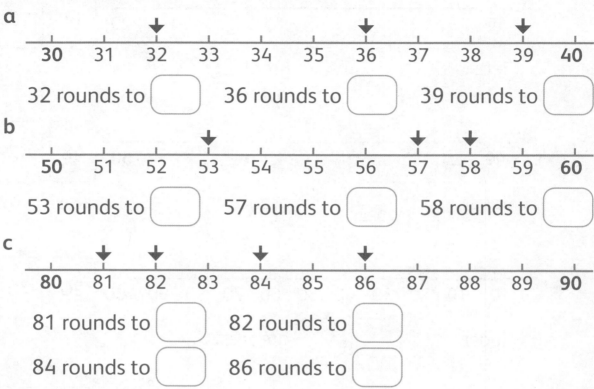

a

30 31 32 33 34 35 36 37 38 39 40

32 rounds to ▢   36 rounds to ▢   39 rounds to ▢

b

50 51 52 53 54 55 56 57 58 59 60

53 rounds to ▢   57 rounds to ▢   58 rounds to ▢

c

80 81 82 83 84 85 86 87 88 89 90

81 rounds to ▢   82 rounds to ▢

84 rounds to ▢   86 rounds to ▢

2 Circle every number that rounds to 70.

60 61 62 63 64 65 66 67 68 69 **70** 71 72 73 74 75 76 77 78 79 80

3 Draw tens and ones to show a number that:

a rounds to 20

b rounds to 30

c rounds to 100

d rounds to 10

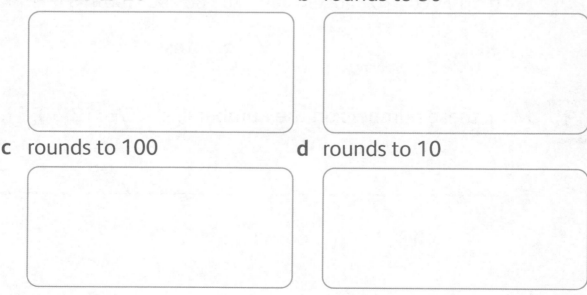

Unit 14　Number patterns and place value

Self-check

 I can do this.

 I can do this, but need to keep trying.

I can't do this yet.

See how much you know!

What can I do?	😉	🙂	🙁
1　I can identify patterns of numbers when counting in ones and twos.			
2　I know what each digit means in 2-digit numbers.			
3　I can decompose 2-digit numbers to identify the place value position of each number.			
4　I can regroup 2-digit numbers in different ways.			
5　I can compare 2-digit numbers using 'greater than' and 'less than'.			
6　I can order a set of numbers on a number line.			
7　I can put objects in order and use ordinal numbers such as 1st, 2nd, and 3rd to identify their positions.			
8　I can round 2-digit numbers to the nearest 10.			

I need more help with:

_____

_____

_____

Unit 15   Addition and subtraction

Can you remember?

20 + 10 = ☐   20 + 20 = ☐   20 + 30 = ☐

50 + 10 = ☐   50 + 20 = ☐   50 + 30 = ☐

Using mental strategies to add and subtract

1 How will you add these small numbers? Write the letter of
the calculation in the Venn diagram to show what you did.

a 6 + 3 + 4
b 3 + 9 + 1
c 6 + 3 + 7 + 6
d 4 + 2 + 1 + 4
e 8 + 4 + 4 + 2
f 1 + 2 + 6 + 6

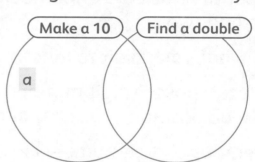

2 Find each answer. Use red to colour in the circles that show
the doubling fact. Use blue to colour in the near double.

a 4 + 5 = ☐   **b** 5 + 6 = ☐   **c** 6 + 7 = ☐

3 Draw jumps of 10 to complete the calculations.

a 28 + 20 = ☐
28

b 28 + 30 = ☐
28

c 38 + 20 = ☐
38

Making estimates

1 The children are working out which two TV programmes they have time to watch. Use estimates to help you.

Sports	Cartoons	Music	History
24 minutes	18 minutes	32 minutes	29 minutes

I have 50 minutes to watch TV. I like history and music. Do I have enough time?

_____

_____

_____

I have 40 minutes to watch TV. Do I have time to watch 2 programmes?

_____

_____

_____

I have 60 minutes to watch TV. Do I have time to watch any 2 programmes?

_____

_____

_____

2 Circle the best estimate each time.

a	22 + 19	Estimate: 50 40 30
b	49 – 22	Estimate: 20 30 40
c	81 – 38	Estimate: 60 50 40

Adding pairs of two-digit numbers

1 Complete these additions on the number line each time.

a 24 + 12 = ☐

b 35 + 22 = ☐

c 46 + 32 = ☐

2 Add each pair of two-digit numbers.

a 22 + 27 = ☐ **b** 45 + 13 = ☐ **c** 64 + 23 = ☐

   32 + 27 = ☐   45 + 23 = ☐   64 + 33 = ☐

3 Draw ones and tens sticks to add each pair of numbers.

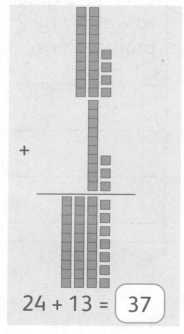

24 + 13 = 37

a

34 + 21 = ☐

b

43 + 32 = ☐

4 Find the totals.

a   32   **b**   33   **c**   45   **d**   45

   + 14   + 16   + 13   + 14

   ☐   ☐   ☐   ☐

5 Make an estimate first. A farmer collects 36 sacks of potatoes on Monday. On Tuesday, he collects 23 more sacks. How many sacks does he collect in total?

My estimate: ☐

☐ sacks in total

Subtracting two-digit numbers

1 Complete each subtraction on the number line.

a 46 – 12 = ☐ |⊥⊥⊥⊥⊥⊥⊥⊥⊥⊥⊥⊥⊥⊥⊥⊥⊥⊥⊥⊥⊥⊥⊥⊥⊥⊥⊥⊥⊥|
46

b 57 – 22 = ☐ |⊥⊥⊥⊥⊥⊥⊥⊥⊥⊥⊥⊥⊥⊥⊥⊥⊥⊥⊥⊥⊥⊥⊥⊥⊥⊥⊥⊥⊥|
57

c 78 – 32 = ☐ |⊥⊥⊥⊥⊥⊥⊥⊥⊥⊥⊥⊥⊥⊥⊥⊥⊥⊥⊥⊥⊥⊥⊥⊥⊥⊥⊥⊥⊥|
78

2 Complete these subtractions.
Draw tens sticks and ones blocks to help you.

a 45 – 13 = ☐

b 38 – 15 = ☐

c 65 – 33 = ☐

d 58 – 35 = ☐

3 Each set of coins shows how much money you have.
Cross out the correct number of coins to show how much is left.
Write the subtraction sentence each time.

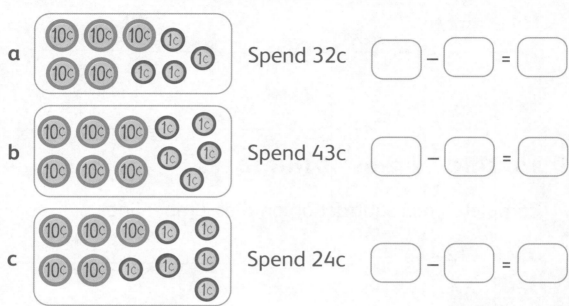

a Spend 32c ☐ – ☐ = ☐

b Spend 43c ☐ – ☐ = ☐

c Spend 24c ☐ – ☐ = ☐

4 Make an estimate first. A farmer grows 95 cabbages.
He picks 23 of them. How many cabbages are left?

My estimate: ☐

☐ cabbages are left.

Addition and subtraction facts to 100

1 Fill in the missing number or numbers each time.

a

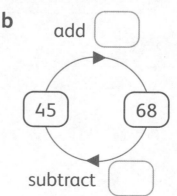

add 14

32   ☐

subtract 14

b

add ☐

45   68

subtract ☐

c

add 24

[ ] 57

subtract 24

d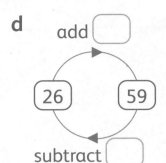

add [ ]

26   59

subtract [ ]

2 Draw jumps on the number line to show the calculation.
Then write the inverse number sentence.

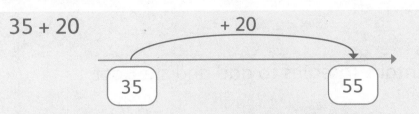

35 + 20

+ 20

35   55

Inverse number sentence: 55 – 20 = 35

a 42 + 30

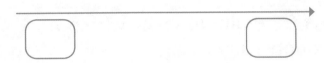

Inverse number sentence: _____

b 67 – 20

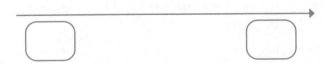

Inverse number sentence: _____

c 56 – 30

Inverse number sentence: _____

Unit 15 Addition and subtraction

Self-check

 I can do this.

 I can do this, but need to keep trying.

 I can't do this yet.

See how much you know!

What can I do?			
1 I can use mental strategies to add and subtract numbers.			
2 I can count on to find the difference between pairs of 2-digit numbers with the same tens value.			
3 I can estimate the answers to additions and subtractions to check the answers.			
4 I can add pairs of 2-digit numbers, with no regrouping through 10.			
5 I can subtract a 2-digit number from a 2-digit number, with no regrouping.			
6 I can show the inverse subtractions for an addition sentence and vice versa, with numbers to 100.			

I need more help with:

_____

_____

_____

Can you remember?

Count on in twos, fives or tens and complete the sentences.

10, 20, ☐, ☐, ☐   5 lots of 10 is ☐

5, 10, ☐, ☐, ☐, ☐   6 lots of 5 is ☐

2, 4, ☐, ☐, ☐, ☐, ☐   7 lots of 2 is ☐

5, 10, ☐, ☐, ☐, ☐, ☐, ☐   8 lots of 5 is ☐

Multiplication as doubling

1 Double the amounts. Write an addition sentence and a multiplication sentence each time.

a

2 litres

☐ + ☐ = ☐ ℓ

☐ × ☐ = ☐ ℓ

b  $10

☐ + ☐ = $ ☐

☐ × ☐ = $ ☐

c  5 kg

☐ + ☐ = ☐ kg

☐ × ☐ = ☐ kg

d  8 cm

☐ + ☐ = ☐ cm

☐ × ☐ = ☐ cm

2 David has more than 3 stickers. Zara has double the amount. Find 5 different ways to make this true.

David					
Zara					

Multiplication tables of 1 and 2

1 Write the multiplication facts to match.

a 1 + 1 + 1 + 1 + 1 + 1

b 2 + 2 + 2 + 2 + 2 + 2

c 1 + 1 + 1 + 1 + 1 + 1 + 1

d 2 + 2 + 2 + 2 + 2 + 2 + 2

e 1 + 1 + 1 + 1 + 1 + 1 + 1 + 1

f 2 + 2 + 2 + 2 + 2 + 2 + 2 + 2

◯ × ◯ = ◯

◯ × ◯ = ◯

◯ × ◯ = ◯

◯ × ◯ = ◯

◯ × ◯ = ◯

◯ × ◯ = ◯

2 Draw jumps on the number line to show:

a 1 × 9 = ◯

```
 |—|—|—|—|—|—|—|—|—|—|—|—|—|—|—|—|—|—|—|—|
 0
```

b 2 × 9 = ◯

```
 |—|—|—|—|—|—|—|—|—|—|—|—|—|—|—|—|—|—|—|—|
 0
```

c 1 × 10 = ◯

```
 |—|—|—|—|—|—|—|—|—|—|—|—|—|—|—|—|—|—|—|—|
 0
```

d 2 × 10 = ◯

```
 |—|—|—|—|—|—|—|—|—|—|—|—|—|—|—|—|—|—|—|—|
 0
```

Multiplication tables of 5 and 10

1

1	2	3	4	5	6	7	8	9	10
11	12	13	14	15	16	17	18	19	20
21	22	23	24	25	26	27	28	29	30
31	32	33	34	35	36	37	38	39	40
41	42	43	44	45	46	47	48	49	50
51	52	53	54	55	56	57	58	59	60
61	62	63	64	65	66	67	68	69	70
71	72	73	74	75	76	77	78	79	80
81	82	83	84	85	86	87	88	89	90
91	92	93	94	95	96	97	98	99	100

a Colour in the numbers in the 5 times table.

b Circle the numbers in the 10 times table.

c What do you notice?

_____

_____

_____

2 Fill in each missing number.

a $2 \times 5 = 10 \times \boxed{}$

b $5 \times 4 = 10 \times \boxed{}$

c $\boxed{} \times 6 = 10 \times 3$

d $\boxed{} \times 8 = 10 \times 4$

e $10 \times 5 = \boxed{} \times 10$

Using arrays

1 A box has 5 pencils in it. Draw an array and write a number sentence for each question.

	Question	Array	Number sentence
	How many pencils in 2 boxes?	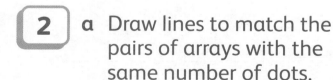	5 × 2 = 10
a	How many pencils in 3 boxes?		
b	How many pencils in 4 boxes?		
c	How many pencils in 5 boxes?		

2 **a** Draw lines to match the pairs of arrays with the same number of dots.

b Now write multiplication sentences for each pair to show that they are equal.

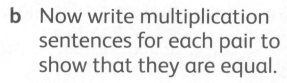

Sharing for division

1 Draw counters to show these divisions as equal sharing.

a 15 ÷ 5 = ☐   b 12 ÷ 2 = ☐   c 20 ÷ 10 = ☐

Grouping for division

1 Write a division number sentence each time.

a The children have 12 shells. They put 2 shells on each sandcastle. How many sandcastles have shells on them?

_____ ☐ sandcastles

b David takes 35 sandwiches to a picnic. He puts 5 sandwiches on each plate. How many plates does he use?

_____ ☐ plates

c Viti washes 60 strawberries. She puts 10 strawberries in each bowl. How many bowls does she use?

_____ ☐ bowls

2 Complete each set of divisions.

a 45 ÷ 5 = ☐   b 60 ÷ 10 = ☐   c 20 ÷ 2 = ☐

40 ÷ 5 = ☐   50 ÷ 10 = ☐   18 ÷ 2 = ☐

35 ÷ 5 = ☐   40 ÷ 10 = ☐   16 ÷ 2 = ☐

Division as repeated subtraction

1 Write a repeated subtraction sentence each time.

a How many groups of 5 pencils are in a pack of 30?

_____ ⬚ groups

b How many plates with 1 sandwich will you get from a tray of 7 sandwiches?

_____ ⬚ plates

c Each ride at the fun park takes 10 people. How many rides are needed for 80 people?

_____ ⬚ rides

2 Fill in the missing numbers. Write the division fact to match.

a

0 ⬚ ⬚ ⬚ ⬚ ⬚ ⬚ 35 40

⬚ ÷ ⬚ = ⬚

b

0 ⬚ ⬚ ⬚ ⬚ ⬚ ⬚ ⬚ 18

⬚ ÷ ⬚ = ⬚

c

0 ⬚ ⬚ ⬚ ⬚ ⬚ ⬚ ⬚ ⬚ 100

⬚ ÷ ⬚ = ⬚

Unit 16　Multiplication and division

Self-check

See how much you know!

 I can do this.

 I can do this, but need to keep trying.

 I can't do this yet.

What can I do?			
1 I can show multiplication as doubles.			
2 I can show multiplication as repeated addition.			
3 I know how to write number sentences for multiplication, using the symbols × and =.			
4 I can recall and use multiplication facts for the 1, 2, 5 and 10 times tables.			
5 I can use arrays to show multiplication.			
6 I know how to write number sentences for division, using the symbols ÷ and =.			
7 I can use sharing to work out a division and explain the method.			
8 I can use grouping to work out a division and explain the method.			
9 I can show division as repeated subtraction.			

I need more help with:

_____

_____

_____

Unit 17   Fractions

Divide to find fractions

1   Use drawings to complete each number sentence.

a $\frac{1}{2}$ of 6 is ☐   **b** $\frac{1}{4}$ of 20 is ☐   **c** $\frac{1}{4}$ of 24 is ☐   **d** $\frac{1}{2}$ of 24 is ☐

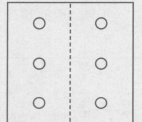 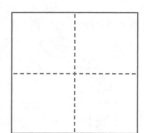 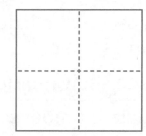 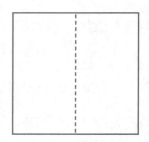

2   David has 14 sweets.

   a   He shares them into halves. How many are in each half?

   b   David wants to share his sweets into quarters. Can he do it? Explain.

3 Complete each fraction sentence.

a $\frac{1}{2}$ of ▢ is 4   **b** $\frac{1}{4}$ of ▢ is 6   **c** $\frac{1}{2}$ of ▢ is 1

d ▢ of 40 is 10   **e** ▢ of 20 is 10   **f** ▢ of 4 is 2

One quarter, two quarters, three quarters …

1 Draw a line to match each fraction to the correct place on the number line.

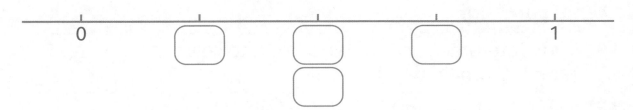

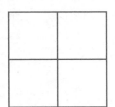 $\frac{1}{2}$  $\frac{1}{4}$  $\frac{3}{4}$ 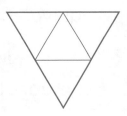 $\frac{2}{4}$

2 Colour in three quarters of each shape.

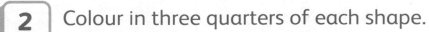

a   b   c   d

3 Find different ways to colour in half of each shape.

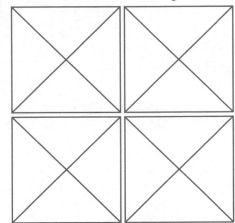   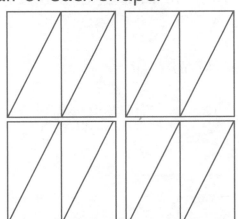

Unit 17 Fractions

Self-check

 I can do this.

 I can do this, but need to keep trying.

 I can't do this yet.

See how much you know!

What can I do?			
1 I can interpret fractions as division to solve practical problems.			
2 I can find one quarter and one half of numbers to 20 that have whole number answers.			
3 I can find one quarter of a shape.			
4 I can read, write and understand $\frac{1}{4}$.			
5 I can use diagrams to show that $\frac{1}{2}$ and $\frac{2}{4}$ are equivalent.			
6 I can show the relative size of $\frac{1}{4}$, $\frac{1}{2}$, $\frac{3}{4}$, and 1 and compare their positions on a number line.			
7 I can use a diagram to explain how combining one half and one quarter makes three quarters.			

I need more help with:

_____

_____

_____

Can you remember?

Complete the table. Fill in the missing months of the year.

January		March	April		June
		September			December

Time

1 Fill in the missing days of the week.

		Tuesday	Wednesday	
		Saturday		

2 Look at Annay's diary to answer the questions below.

Monday dentist 4:00 p.m.
Tuesday swimming lesson 11:00 a.m.
Wednesday
Thursday
Friday Guss coming to tea
Saturday
Sunday Football 3 p.m.

a On which day is Annay going to the dentist?

_____

b On which day is Guss coming to tea?

_____

c On which day does Annay have a swimming lesson?

_____

3 Jack visited Grandmother on Monday. He stayed for 3 days.

On which day did Jack return home? _____

4 It is Saturday. Elok's piano test is in 4 days' time.

On which day is Elok's piano test? _____

5 Viti posted a parcel on Wednesday. It took 2 days to arrive.

On which day did the parcel arrive? _____

Capacity

1 Estimate the capacity of these containers. Draw lines to match.

a

| 10 ℓ |
| 100 ℓ |
| 250 ml |
| 2 ℓ |

b

c

d

2 Use <, > or = to compare the amounts.

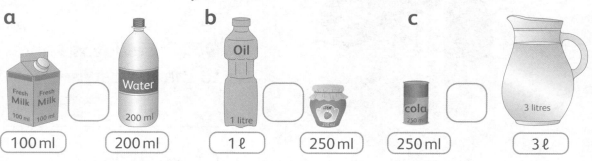

a | 100 ml | 200 ml   b | 1 ℓ | 250 ml   c | 250 ml | 3 ℓ

3 How much water is in each measuring jug?

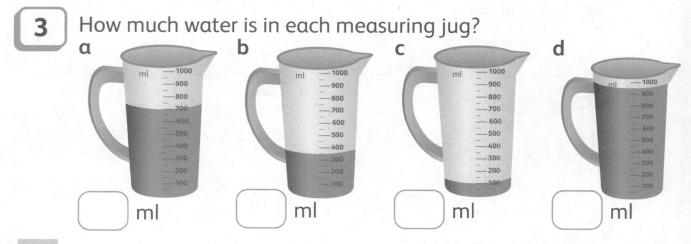

a ( ) ml   b ( ) ml   c ( ) ml   d ( ) ml

Measures

1 Zara is measuring the length of her book. Which instrument should she use?

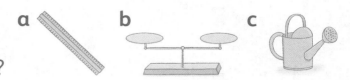

2 Annay is baking some bread. Which measuring instrument should he use to weigh the flour?

3 Elok is measuring some milk. Which instrument should she use?

4 How long is each piece of string?

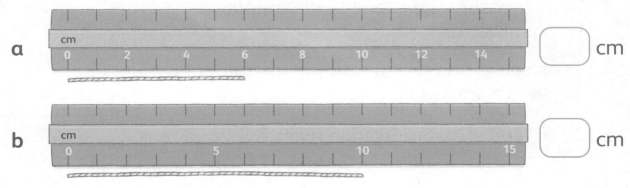

a _____ cm

b _____ cm

5 How heavy is each vegetable?

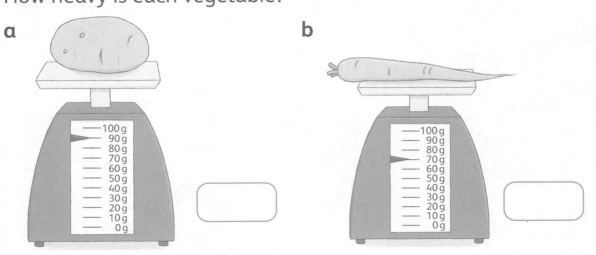

a

b

Unit 18 Time and measurement

Self-check

 I can do this.

 I can do this, but need to keep trying.

 I can't do this yet.

See how much you know!

What can I do?			
1 I know that calendars are used to organise time in days, weeks and months.			
2 I can estimate the capacity of a container before measuring.			
3 I can measure the capacity of containers using millilitres.			
4 I can select and use measuring instruments for length, mass and capacity.			
5 I can read scales marked in twos, fives and tens.			

I need more help with:

_____

_____

_____